Lifeboat Design and Development No.7

A CLASS
INSHORE LIFEBOATS

The RNLI's A class rigid-hulled inshore lifeboats, their design and history

Nicholas Leach

FOXGLOVE PUBLISHING

First published 2021

Published by
Foxglove Publishing Ltd
Foxglove House, Shute Hill,
Lichfield WS13 8DB
United Kingdom
Tel 07940 905046

British Library Cataloguing in Publication Data.

ISBN 9781909540224

Typesetting/layout by
Nicholas Leach/
Foxglove Publishing

LIFEBOAT DESIGN AND DEVELOPMENT • ALREADY PUBLISHED

LIFEBOAT DESIGN AND DEVELOPMENT SERIES This is the seventh book in a series of concise illustrated volumes tracing the history and describing technical aspects of different RNLI motor and inshore lifeboat types. The volumes in the series published so far are as follows: No.1 Clyde recuse cruisers; No.2 Surf motor lifeboats; No.3 Atlantic inshore lifeboats; No.4 Tyne fast slipway lifeboats; No.5 Barnett motor lifeboats; and No.6 Brede intermediate lifeboats.

THE AUTHOR Nicholas Leach has a long-standing interest in lifeboats and the lifeboat service. He has written many articles, books and papers on the subject, including a history of the origins of the lifeboat service; a comprehensive record of the RNLI's lifeboat stations in 1999, the organisation's 175th anniversary; RNLI Motor Lifeboats, a detailed history of the development of powered lifeboats; and numerous station histories, including ones covering the stations of Cromer, Longhope, Padstow, Sennen Cove, Weymouth and Humber. He has visited all of the lifeboat stations in the UK and Ireland, past and present, and is Editor of Ships Monthly, the international shipping magazine, and Lifeboats Past & Present, the magazine for lifeboat enthusiasts.

Contents

Acknowledgements

A number of people have assisted with this publication, and I am grateful to them all. At the RNLI, Hayley Whiting of the RNLI Heritage Trust provided research facilities and answered various queries. Many people assisted with photographs and information: in particular Simon Culliford at Falmouth; Nick White, LOM at Littlehampton; Leonie Mead, Oban Lifeboat Press Officer; Kevin Dent, of Invergordon; and Clive Lawford. Richie Leonard, LBES Archivist, and Scott Snowling supplied extra information and data; and Ian Moignard commented on the final text with thoroughness and attention to detail.

Nicholas Leach
Lichfield, November 2021

Bibliography

Fry, Eric C. (1975): Life-boat Design and Development (David & Charles, London).

Hawkes, Andrew (1995): Lifeboatmen Never Turn Back: Poole Lifeboat Service from 1826 (Poole Historical Trust).

Hatch, G.N. (1967): Fast Inshore Rescue/Boarding Boats (Paper No.9 in International Lifeboat Conference Report, France).

Horobin, Richard (2017): Lyme Regis Lifeboat Station: The Pearl of Dorset (RNLI).

Leach, Nicholas (1992): The.

— (2005): RNLI Motor Lifeboats (Landmark Publishing Limited, Ashbourne).

Salsbury, Alan (2002): A History of the Torbay Lifeboats: With Pride and Dedication (Halsgrove Publishing, Tiverton, Devon).

Summary of the A class inshore lifeboats

ON	Year Type	Stations (launches/lives saved) Notes
A-1 (ex 18-01)	1966 Hatch	Displayed at International LB Conference, St Malo, France, 6.1967; Falmouth 1967-67, Shoreham 1967-67, Plymouth 7.1968-5.72
A-2 (ex 18-03)	1966 Hatch	Poole 1967-69, Torbay 7.1969-10.1974
A-500 (ex 17-01)	1968 Dell Quay Dory	Eastney 8.1968-6.1969, Ramsgate 7.1969-12.1971 (23/16), Poole 2.1972-12.1973 (21/7)
A-501 (ex 17-002)	1969 Dell Quay Dory	Lyme Regis 6.1969-9.1973 (51/49), Poole 7.1974-5.1975 (26/7)
A-502 (ex 17-003)	1969 Dell Quay Dory	Poole 6.1969-2.1972 (28/19), Ramsgate 2.1972-5.1975 (28/7), Poole 8.1975-1.85 (275/38)
A-503 (ex 18-02)	1967 McLachlan	Eastney 7.1967-7.70 (7/6), Poole 1971 (8/0), Weston-super-Mare 1971 (5/0), Pill 7.1971-10.72 (25/6)
A-504 (ex 18-004)	1970 McLachlan	Weston-super-Mare 5.1970-12.1983 (171/60)
A-505 (ex 18-005)	1970 McLachlan	Eastney 7.1970-9.71, Oban 5.1972-5.1973 (12/6), Eastney 12.1974-10.75 (30/6), Humber (as Boarding Boat) 1977-87
A-506 (ex 18-006)	1971 McLachlan	Peel 5.1972-6.1976 (3/3), Poole 12.1976-8.79 (1/0), Plymouth 8.1979-8.80 (20/0)
A-507 (ex 18-007)	1971 McLachlan	Weston-super-Mare 12.1972-2.73 (3/0), Peel 5.1973-8.76 (15/8) , Ramsgate 1977 (1/0), Plymouth 1977-10.83 (58/12)
A-508 (ex 18-008)	1971 McLachlan	Eastney 9.1971-12.1974 (41/26), Invergordon 7.1976-11.79 (1/0), Falmouth 4.1980-1.1988 (92/19), Sold 1988
A-509	1972 McLachlan	Plymouth 5.1972-10.76 (85/22), Brighton 5.1978-7.78 (2/0), Falmouth 1981-83 (5/10), Sold 1988
A-510	1973 McLachlan	Pill 6.1973-10.1974 (14/10), Ramsgate 6.1975-8.1984 (141/32), Falmouth 1985-86 (3/0), Humber (boarding boat) 6.1986-1.87
A-511	1973 McLachlan	Oban 6.1974-1982 (87/9), Falmouth 4.1984-9.1985 (15/20) Sold 9.1985
A-512	1974 McLachlan	Torbay 3.1975-6.1987 (176/55), Falmouth 1-5.1988 (0/0) Scrapped at Grimsby 5.1988
A-513	1984 Boston Whaler	Poole 1.1985-12.1994 (592/106), capsized on service 23.11.1986, removed from station for repair, returned 10.6.1988

Two of the boats were named: A-501 was named Bob Abbott (1969-73) in memory of Coxswain Robert W. Abbott, from Lyme Regis; A-513 was named Sam and Iris Coles.

Introduction

The rigid-hulled inshore lifeboats introduced by the Royal National Lifeboat Institution (RNLI) were an unusual development and should be examined in the context of lifeboat design in the 1960s, when the RNLI was looking at a wide range of new rescue boat designs and different construction materials, making assessments about which would be most suitable to meet the changing demands being made on the charity as more people went to sea for leisure purposes. The wooden motor lifeboats, which were largely based on pre-war designs, were slow and unsuited to the rescues being undertaken in the post-war era. The inflatable inshore rescue boat, introduced in 1963, went some way to meeting these new demands, but within a few years the RNLI was looking at inshore rescue craft which were more capable and faster.

At the same time, the RNLI had been investigating the possible use of glass reinforced plastic (GRP) construction for lifeboats, and in

▼ The 18ft 6in McLachlan A-508 in service at Falmouth, the station she served for eight years during the 1980s. (By courtesy of Falmouth RNLI)

▲ The 18ft 6in McLachlan A-510 moored at Ramsgate. (Clive Lawford)

fact the rigid-hulled ILBs, later designated the A class when ILBs were reclassified, were the first lifeboats to be built of GRP. The experimental prototypes of two designs, the Hatch and McLachlan, named after their designers, were built in wood, but when produced in quantity the intention was that they would have GRP hulls.

Only two Hatch boats were built, but ten McLachlans saw service. In addition, three other A class boats were built by the 1960s by Dell Quay Productions Ltd, based on its commercially available Dory design. The other A class inshore lifeboat, a Boston Whaler Outrage, was a one-off built specifically for service at Poole, lasting into the 1990s and proving to be the last of her kind. The rigid-hulled ILBs were primarily intended to serve at stations where there was a need for a rescue craft faster than a conventional lifeboat and with a greater range than an inflatable rescue craft, but they were operated at stations on their own and were well liked by their crews. However, by the mid-1970s the Atlantic rigid-inflatable was deemed to be a more suitable rescue craft, and has subsequently been built in large numbers. The various A class designs can therefore be seen as something of an anomaly in the history of lifeboat design.

A class ILB design and development

During the 1960s, the RNLI assessed various ways of improving the effectiveness of the inflatable inshore rescue boat (later inshore lifeboat (ILB), and designated the D class in 1972) after it had become part of the fleet during the early years of the decade, being well established around the coast by the 1970s. ILBs had been introduced in response to an increasing number of incidents to which conventional offshore lifeboats were being called but which required a fast response and usually occurred in relatively benign conditions. More people were using the sea for leisure in the 1950s and 1960s, as incomes and leisure time increased and, as a result, the number of inshore incidents increased. The conventional lifeboats, with a top speed of nine knots at best, were not ideal for such work and a simple, fast rescue craft was required. The inshore rescue boat, in the form of a small single-

◀ The first inshore lifeboats, such as this at Aberystwyth, were inflatables powered by a single engine, and were deemed suitable for daylight operations in moderate conditions. (By courtesy of the RNLI)

▲ The 18ft 6in McLachlan 18-005 moored at the Eastney side of the Hayling Ferry pontoon, at the entrance to Langstone Harbour, while she was on station at Eastney in the 1970s. Two experimental Atlantic rigid-inflatables are moored alongside: C-1 on the outside with, probably, B-1 in the centre. The RNLI trialled a series of inshore lifeboats during the 1970s. (Grahame Farr, by courtesy of the RNLI)

engined inflatable, was the answer.

The inflatable ILB soon proved its worth after ten were introduced at selected stations around the country in 1963. The 16ft inflatables, made from tough nylon with neoprene, were crewed by two or three, powered by a 40hp outboard engine, and could be launched quickly. Their advantages were their speed which, at twenty knots, was faster than any lifeboat then in service; their ability to go alongside other craft or persons in the water without causing or suffering damage; the short time taken to launch them; and the relatively low construction and running costs.

Such was their success that, in each of the years after the first inflatables had gone to their stations, more places began to operate them. By 1966 inflatables were operating at seventy-two stations, of which thirty-two remained in service throughout the year, with the remainder operating only during the summer. Some were co-located with offshore lifeboats, such as at Gorleston, and others operated on their own, such as at Whitstable. As their use became more widespread and their effectiveness as rescue craft more evident, the idea to produce an inshore boat with greater capabilities, including night operations, was considered, and to this end a number of different types were assessed.

A whole series of inshore craft were designed and developed during the late 1960s and 1970s, of which the rigid-hulled A class, the subject of this book, was just one. At the same time as the A class was being developed, so was the rigid-inflatable B class, which was enthusiastically accepted by crews who had it for evaluation in 1970. During the next few years the rigid-inflatable, designated the Atlantic 21, gained widespread acceptance. The Atlantic proved to be the most suitable and was subsequently deployed throughout the UK and Ireland (see Lifeboat Design and Development No.3). There were also twin-engined inflatables, designated the C class, as well as Atlantics of different lengths, before the 21ft version became the standard one in service.

The rigid-hulled ILBs covered by this book were designated the A class from 1972. A total of sixteen were built, the great majority in the late 1960s and early 1970s. At this time standardisation of lifeboat types within the RNLI fleet was not a priority, and consequently various different types and designs were utilised and operated. Small numbers of bespoke craft were often the order of the day, with boats being designed for service at specific stations, to contend with specific local conditions.

The need for larger inshore craft led to the development of two different rigid hull designs: the Hatch and the McLachlan, named after their respective designers. They were slightly larger than the inflatable, and were regarded as being suitable in filling the gap between the inflatable boats costing about £1,000 each and the relatively heavy offshore lifeboats, such as the 48ft 6in steel-hulled Solent, which then cost upwards of £45,000 each. They were also intended to have a role as a boarding boat,

◀ The experimental inshore rescue boat X.5 at Lyme Regis for evaluation trials. The boat was one of several built at Atlantic College in Wales. (Grahame Farr, by courtesy of the RNLI)

to take lifeboat crews to lifeboats at moorings. Experience had shown that boarding boats were quite often needed for rescues when speed was essential, and the newly-designed rigid-hulled boats, faster than any other lifeboat then in service, therefore fulfilled this need.

The prototype Hatch and McLachlan boats were used largely for evaluation purposes and they were built of wood. However, it was intended that glass reinforced plastic (GRP) would be used for hull construction when the design went into production. This would, it was hoped, reduce both the initial cost and maintenance, provided the abrasion resistance of the material, then in its infancy, had been sufficiently proven. Ultimately, only two Hatch boats were built, both of wood, but for the small production run of McLachlans the later boats were moulded from GRP.

A third type of rigid-hulled boat was also acquired, but from a commercial manufacturer. This was the Dell Quay Dory, of which three were built for the RNLI, each with a GRP hull, and powered by twin 40hp outboard engines. They had a top speed of twenty-five knots and an endurance of three hours, but were not intended for use in exposed sea conditions. They were primarily intended to lie afloat at moorings, but could also be used in a dual role as a boarding boat, as described above. After the initial Dorys had entered RNLI service, no more were acquired.

▼ The prototype Hatch boat, originally given the operational number 18-01, pictured on trials. It was essentially an experimental craft and was unlike any other lifeboat design. (By courtesy of the RNLI)

▲ The first Hatch boat, 18-01, on trials. (By courtesy of the RNLI)

The Hatch boats

The first rigid-hulled craft tested by the RNLI was designed by and named after George N. Hatch, AMRINA, a senior draughtsman with the RNLI. The design requirements stated that the boat must have a speed well in excess of twenty knots, be 18-20ft in length, unsinkable with eight people on board, capable of sustained high speed in rough water (with a possible limit of force eight) and must be of shallow draught.

Hatch designed a boat which was 20ft 6in overall, had a beam of 7ft 3in and an overall depth of 4ft 3in. The high deadrise hull had double chines aft and a wide foredeck flared out from the deep-vee forward sections. The hull was similar in form to that of offshore power boats, with a moderately high freeboard and a flared bow, and in calm conditions speeds of up to twenty-six knots were achieved. The hull was divided by watertight bulkheads to give four watertight compartments. A large open self-bailing cockpit provided sufficient space for at least eight people to be taken on board. The boat had wide flush decks fore and aft, with handrails and lifelines. Foamed polystyrene blocks were fitted throughout, including in the fore peak, the after compartments, below the cockpit sole and under the side decks. The blocks were made portable to allow access to the bilges and hull for inspection and repair.

The crew were provided with two padded seats, vertically sprung, hydraulically damped and adjustable fore and aft, with quick release

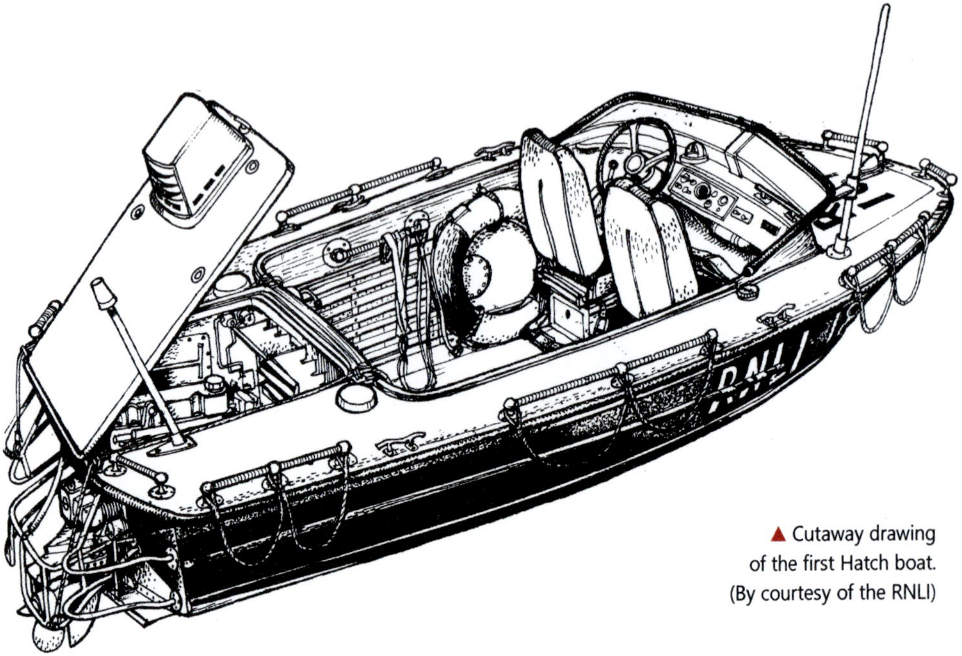

▲ Cutaway drawing of the first Hatch boat. (By courtesy of the RNLI)

nylon webbing seat belts and a tubular footrest with nylon webbing toe straps. This meant the helm was free to manipulate the controls and the second crew member could operate the radio. The crew gear was a nylon coverall suit, inflatable Lifemaster life-jacket, safety harness for use when moving about the craft, and crash helmet with visor. A windscreen was not fitted as it was considered a better compromise to allow easy access to the fore deck direct from the cockpit, and protection from the elements was provided by the helmet, visor and coverall suit.

An inboard engine with outboard drive was fitted, providing the control and installation advantages of outboard drives, without the vulnerability of the outboard engine. The prototype was fitted with a Volvo-Penta Aquamatic AQD.110 petrol engine coupled to a Type 100 outdrive. The engine was rated at 110bhp at 5,100rpm giving approximately 95shp, and was driven through 1.66:1 bevel gear train, coupled to a three-bladed propeller 13 inches in diameter by 13 inches pitch.

The engine was placed in a totally enclosed space covered by a full length watertight hatch, hinged at the aft end. Air was taken in through a large capacity water-excluding vent box coupled to a vent trunk at the bottom of the engine space, so the compartment was ventilated while the engine was running. To protect the outdrive unit from damage, and to

◀ The first Hatch boat, with operational number 18-01, as she was configured when first built, being demonstrated as a boarding boat. (By courtesy of the RNLI)

give a measure of protection to survivors against possible injury from the propeller, tubular galvanised steel stern-guards were fitted to the transom. The centre section was arranged to hinge up under the deck overhang.

The hull was capable of being driven moderately fast in rough seas and, in more testing conditions, at reduced speed. The initial sea trials saw the boat taken out in a force eight gale, with waves up to about 8ft in height with breaking crests, and force five to six winds in very confused seas of 4-6ft in height. The boat was able to make good progress despite the challenging conditions. Another trial involved continual and rapid throttle and helm changes, during which the helmsman maintained

▼ Hatch boat 18-01 after she had been modified for self-righting in 1968 with the addition of buoyancy blocks at bow and stern. These blocks were subsequently removed (see p.33). (By courtesy of the RNLI)

control. The brief initial trials proved the adequacy of the hull form and crew seating arrangements, which enabled prolonged searches to be made without tiring the crew, a feature lacking in the inflatable boats.

Two Hatch boats were built by William Osborne Ltd at Littlehampton in 1966. The main difference between the boats was the engine: 18-01 had a petrol engine, 18-03 had a diesel. They were originally numbered 18-01 and 18-03, but were later designed A-1 and A-2 after the RNLI changed its numbering system for inshore lifeboats. The 'A-' designated a rigid hull; other types in the inshore fleet were designated 'B-' for rigid inflatables (the Atlantic class) and 'D-' for the fully inflatable.

The first Hatch boat was displayed at the International Boat Show at Earls Court in January 1967, before being sent to Shoreham Harbour and Falmouth for trials later in the year, and in 1968 went on station at Plymouth. The second boat, 18-03, went to Poole in July 1967, staying for two years before undertaking a five-year stint at Torbay. Only two Hatch boats were built as, during 1968, while the boat was being trialled, the RNLI deemed its other fast rigid-hulled boat for inshore rescue, the McLachlan, to be superior and so pursued this design instead.

Despite being experimental craft, both Hatch boats saw operational service, with the second, 18-03, being involved in a particularly fine service while she was at Torbay. On the afternoon of 5 October 1973 the police received a report that a woman was in the sea at Meadfoot Beach, Torquay. A south-westerly force seven wind was blowing across Tor Bay, accompanied by rough seas. 18-03 left her moorings at 3.50pm crewed by Coxswain Kenneth Gibbs and Mechanic Barry Pike. At Meadfoot Beach, the seas were breaking among the rocks and seaweed, producing waves of 8-10ft in height. Coxswain Gibbs commenced a search of the beach, being aware that the heavy waves could drive the wooden hull of his boat onto one of the submerged rocks. But with great skill and seamanship, he held the Hatch boat just inside the line of the breaking waves. Barry Pike soon spotted the woman among the breakers and, without thought for his own safety, dived into the surging waves in an attempt to rescue the casualty. As he did so, a large wave approached the ILB and Coxswain Gibbs, fearing for the safety of his colleague and the casualty, had to use all his seamanship skills to keep the ILB from being swept onto the two people in the water, one of whom was his own crew.

▲ Hatch boat 18-03 on trials. Unlike 18-01, this boat was not altered during the course of her RNLI career. (By courtesy of the RNLI)

Barry later recalled: 'I saw the girl and made a grab for her, but I was thrown over and over, in a cartwheel, by the waves. I think I had hold of her at one point, but as I tried to stand up I lost my footing because of the undertow and the shingle being swept away from under me.' Becoming exhausted by his efforts, Barry was washed up onto the beach. Regaining his breath, he again entered the water but, as he was so exhausted, he was beaten back by the breakers and had to be pulled from the surf by Doug Trethowan, a Police Constable at Torquay, who was at the scene. Barry tried again to reach the woman, and eventually grabbed hold of her, with the policeman pulling him to safety. Barry recalled: 'I was thrown up on the beach at one stage [but] . . . I didn't realise the difficulty I was in, but the Policeman threw me a line and actually saved my life.'

The Constable, P.C. Trethowan, said afterwards: 'I was on patrol and saw a group of people looking out to sea. It was a lousy day and blowing a

◀ 18-03 in Brixham harbour on station at Torbay. (By courtesy of the RNLI)

Hatch and McLachlan principal particulars

	HATCH	McLACHLAN
Length	20ft 6in (6.27m) overall, 18ft 6in (5.65m) moulded, 16ft 8in at rest (5.71m) waterline	18ft 6in (5.65m) overall, 16ft 4in (5.9m) waterline
Beam	7ft 2in (2.18m) , 6ft 11½in (2.12m) moulded	8ft 3in (2.45m)
Depth	3ft 1½in (0.95m) moulded	
Displacement	2,665lbs (1,211kg) with full fuel and 2 crew	Approx 3,000lbs (1,362kg)
Engines	18-01: Volvo-Penta Aquamatic AQD.110 petrol engine, rated at 110bhp, 5,000rpm 18-03: diesel engine	First four boats: twin 60hp Ford Series 2255E petrol engines, marinised by Watermota Ltd, Perkins outboard Z drives
Speed	25.75 knots maximum continuous	22 knots maximum
Fuel capacity	24 gallons (109 litres) (2 x 12 gallon tanks)	30 gallons (136 litres)
Fuel consumption	4.4 gallons/hr (20 litres/hr) at 5,000rpm	

gale; waves were crashing against the sea wall and breaking over onto the road. I was told that there was a woman in the water. I saw the inshore lifeboat at the edge of the waves and I saw a lifeboat man dive into the water in an attempt to rescue her. The lifeboatman was washed into the beach and then out again. I could see he was in difficulty in the surf so I went back to the van and got the tracking harness with a 32ft lead. I waded out as far as I could but the waves were breaking over me. Luckily Barry reached out and got hold of the harness and I was able to pull him in. He went out again to rescue the woman and luckily I got him back with the harness.' Regrettably the life of the woman could not be saved. However, in recognition of his outstanding courage and bravery, the RNLI awarded Barry Pike a Silver medal, and the Thanks Inscribed on Vellum were accorded to Coxswain Ken Gibbs.

The McLachlan boats

Designed by and named after J.A. McLachlan, MRINA, of G.L. Watson & Co of Glasgow, the second of the fast rigid-hulled inshore lifeboats the RNLI developed during the late 1960s incorporated several notable features, including a ragged chine hull and seawater ballast stability tanks. The ragged chines were a series of steps which effectively reduced the amount of the hull's surface which was in contact with the water, and thus minimised the pounding in normal hard-chine boats. Nine chines extended the full length of the hull, each inclined at an angle to the water to cushion the underside and deflect spray.

◀ The first McLachlan lifeboat, 18-02, at Cowes shortly after being completed. (By courtesy of the RNLI)

The first of the McLachlans, designated 18-02, was 18ft 6in in length and 8ft in beam. It was built by the Cowes shipyard of W.A. Souter at the same time as the Hatch boat was under construction at Littlehampton. To achieve maximum strength with minimum weight, the hull was moulded using three skins of mahogany, which was the builder's standard practice for high-speed craft. It was subdivided into several watertight compartments, each filled with polyurethane foam, comprising: fore peak, aft wing peak side decks, and beneath cockpit deck, giving, what was described in the RNLI's contemporary literature,

▼ The first McLachlan lifeboat 18-02 on trials. (By courtesy of Simon Culliford)

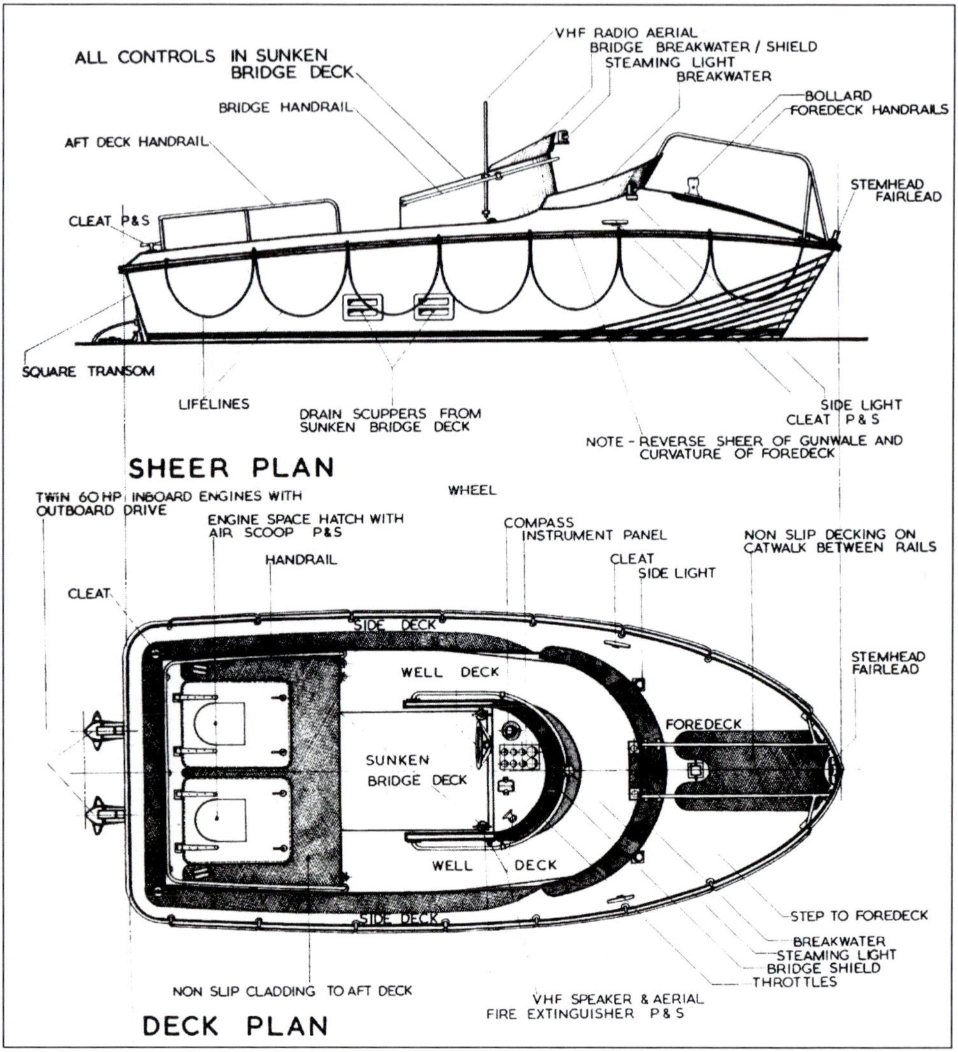

▲ Diagram of the 18ft 6in McLachlan from Eric Fry's Lifeboat Design and Development. The standard RNLI-pattern side scuppers in the hull provided adequate drainage.

'an ample excess of positive buoyancy'. The three water stability ballast tanks were positioned forward, centre and aft.

The steering and engine controls were located amidships within a semi-enclosed streamlined bridge structure, which housed the wheel, compass, engine throttles and radio equipment. This position offered the helm unrestricted movement and provided good all-round visibility. Draining was achieved through the use of standard RNLI-pattern side scuppers. The high foredeck provided reasonable protection for at least eight survivors, while the crew of two were sheltered by the midships structure, which had its own breakwater and wind deflector.

A single 90bhp Evinrude stern drive was installed initially in 18-02. This gave a high power-to-weight ratio as well as flexibility of control and good manoeuvrability in most sea conditions, and enabling the boat to achieve a speed of 20.9 knots in calm water. The deep-vee hull section and the boat's relatively short length were critical factors in the weight and speed, and resulted in the speed being considerably reduced when three or four extra people were on board. In the early 1970s 18-02 was re-engined and modified so that she could take twin Perkins 4.108m diesel engines, each of 47hp.

Leaflet No. 28

Royal National Life-boat Institution **RNLI**

NEW RESCUE/BOARDING BOAT

An interesting feature of this new rescue/boarding boat is the positioning of the steering and engine remote controls in a semi-enclosed streamlined structure amidships.

▲ RNLI Leaflet No.28 published in the late 1960s described the McLachlan boat.

Following the trials with 18-02, in September 1969 the RNLI ordered a further four McLachlans with GRP hulls from William Osborne, at Littlehampton, at a cost of approximately £3,400 each. They were built under the supervision of the designer, J.A. McLachlan, and entered

◀ 18-02 on trials in the 1960s shortly after she had been completed. Originally fitted with a single engine, she was later modified considerably and fitted with twin outboard engines, All later McLachlans had twin engines. (By courtesy of the RNLI)

▶ 18-02 on display at the London Boat Show at Earl's Court. (By courtesy of the RNLI)

▲ McLachlan ILBs 18-02 and 18-004. The latter, later renumbered A-504, was the first glass-reinforced plastic-hulled boat and served at Weston-super-Mare from 1970 to 1983. (By courtesy of the RNLI)

service in the early 1970s. The four production boats, and indeed all subsequent McLachlans, measured 18ft 6in in length, 16ft 4in at the waterline, and had a beam of 8ft 3in. Numbered 18-004 through to 18-007, they were fitted with twin 60hp Ford Series 2255E petrol engines marinised by Watermota Ltd, and Perkins 'Z-drive' units were used.

18-02 was used for trials at a number of stations, and in July 1967 she was placed on operational duty at Eastney (now known as Portsmouth) lifeboat station, covering the busy waters off the Hampshire coast and in Langstone harbour. She was followed into service by the four production boats, the first of which went to Weston-super-Mare, where it was well liked. Another five McLachlans were then built, all by William Osborne

18-004, later renumbered A-504, served at Weston-super-Mare, where she spent thirteen years covering the dangerous waters of the Bristol Channel. (By courtesy of the RNLI)

at Littlehampton, in the early 1970s, with the last, numbered A-512, entering service at Torbay in early 1975. Most of the McLachlans were kept afloat at moorings, being used as both boarding boats and inshore lifeboats at stations such as Falmouth, Ramsgate and Torbay. However, the boat at Weston-super-Mare, 18-004 (later A-504), was launched from a slipway using a special carriage and another, A-507 at Peel, was launched from a carriage towed by a tractor down a concrete ramp.

While the McLachlan boats were used in service with the standard configurations, A-512, which served at Torbay for twelve years, was modified in 1977-78 by the removal of the standard steering console, and two forward-facing seats were fitted, together with a windscreen and a dashboard for the steering and engine controls. She remained in

A-507 at Peel on her launching trailer, being pulled out of the lifeboat house. The tractor T43 (JXR 934) was built in 1948 for launching offshore lifeboats, and was at Peel for five years in the 1970s.

▶ McLachlan A-510 making her first run after her arrival at Pill on 22 July 1973. She was fitted with a reinforcing shoe on her forefoot to combat wear when she came into contact with the ferry slip at Pill. After leaving Pill, she went to Ramsgate, where she served for nine years. (Grahame Farr, by courtesy of the RNLI)

service at Torbay until January 1988, when she was moved to Falmouth. After three months at Falmouth she was withdrawn in late April 1988, becoming the last McLachlan to be taken out of service, and ending a chapter in the development of small fast rigid-hulled lifeboats.

The McLachlans were involved in many rescues, none more noteworthy than that carried out by A-504 at Weston-super-Mare on 13 September 1975. She was launched just after 10pm, crewed by Helm Julian Morris, Bernard Watts and Ian Watts, in a north-westerly force nine gale and rough seas. Within ten minutes the crew sighted a flashing light close inshore, and found a local motor boat stranded on rocks. Seas were breaking over the boat, which was ashore in a small cove bounded

▶ A-512 was the last McLachlan to be built, and served at Torbay for twelve years. She was modified during her time in service, with the removal of the centre steering console. She was the only McLachlan to which this modification was made.

▲ A-510 returning to Ramsgate, where she was stationed for almost ten years. (By courtesy of Ray Noble)

◀ A-511 at Oban in July 1974 with the Moderator of the General Assembly of the Church of Scotland, the Rt Rev Dr David Steel, on board during a publicity visit. (By courtesy of the RNLI)

▲ A-508 and A-507 at the RNLI Depot, Poole in July 1984. They were sold out of service in 1988 and 1989 respectively. (Nicholas Leach)

by sheer cliffs that offered no escape for the five people, and the tide was rising. Helmsman Morris took the ILB to within ten yards of the casualty despite the rough and confused seas, and the submerged rocks.

He then took the ILB out to sea, anchored about thirty yards off the casualty, and veered down with both engines raised but running. The first attempt failed, but on the second attempt the ILB, despite striking the rocks, was veered to within 20ft of the casualty and a line thrown ashore. This was secured around the waist of a survivor and he was pulled through the water to the ILB. Four men were rescued in this way; the fifth, the owner of the boat, managed to wade out to the ILB.

All the survivors were safely on board by 11.05pm, and they were landed at the Ferry Stage, Uphill. Back at Weston, difficulty was experienced in recovering the ILB because of the very rough weather and it was unsafe for the helpers to stand on the slip, but the ILB was eventually taken out of the water just after midnight. For this outstanding service the Bronze medal was awarded to Julian Morris, and the Thanks of the Institution inscribed on vellum were accorded to Bernard Watts and Ian Watts.

Dell Quay Dory fast rescue boats

As well as designing the McLachlan and Hatch boats in the late 1960s, the RNLI also assessed a commercially designed rigid-hulled boat, known as a Dory, built by Dell Quay Productions Ltd at Itchenor, Sussex. The Dell Quay Dory was built of glass reinforced plastic (GRP), with the cavity between hull and deck filled under pressure by rigid polyurethane foam. The standard hull was strengthened for RNLI service so that it could take an aluminium steering console. Although the intention was

that, if the boat went into production, a GRP moulding would be used for the steering console, as it turned out, only three of these boats were used by the RNLI. Within the hull were two eighteen-gallon petrol tanks and a locker to store equipment. As with the Hatch and McLachlan, the Dell Quay Dory was intended to serve as both a boarding boat and a high-speed rescue craft.

▲ The RNLI's first Dell Quay Dory 17-01 on trials, fitted with Evinrude outboards. (By courtesy of the RNLI)

The Dory measured 17ft in length and 7ft in beam. The draught with engines in use was just eighteen inches, and only eight inches with the engines raised, and this very shallow draught was one of the design's main benefits. The transom had a solid marine ply core bonded to the hull and deck laminates. The first boats had two 36hp Penta outboard engines to give a top speed of twenty-five knots, but a variety of engines were subsequently fitted. Each engine had a separate steering cable, the engines only being connected at the wheel. This enabled the boat to

Dell Quay Dory principal particulars

Dimensions	Length 17ft, beam 7ft, draught 8in (motors raised); 18in (motors lowered)
Displacement	1,030lb; integral reserve buoyancy 3,000lb
Engines	Twin outboards of various makes, including Penta, Mercury and Evinrude
Speed	Maximum 25 knots (with two crew); 15 knots on one engine
Fuel capacity	36 gallons
Fuel consumption	8 gallons per hour (estimated); range 110 miles (estimated)

be run with one engine tilted, so that in the event of damage or failure of one engine, the performance was not affected too adversely. The steering gear was by Morse and the engines controlled via a Morse single lever. The engines had manual starting, as the complication of electric starting was deemed unnecessary.

The first of the Dell Quay Dory boats, 17-01, was notable in that it was the first rescue boat with a GRP hull to carry out operational trials in RNLI service. The boat was demonstrated to the press at Warsash on 19 August 1968. Among those who attended this showing was Grahame Farr, the lifeboat author and historian, who wrote: 'The speed and manoeuvrability of the Dory, powered by twin Evinrude B.16 40hp outboards, were impressive. Her steadiness was also remarkable and on several occasions the unusual construction of the hull, with deep chines forming virtually second and third keels, was clearly shown as she leaped over the crests of the waves in our wake.'

17-01 was temporarily stationed at Eastney, Hampshire after the press announcement, taking up her duties later in August 1968. She stayed there for less than a year before being moved to Ramsgate and then subsequently went on to Poole. Two more Dell Quay Dorys were completed for the RNLI in 1969, but after that no more were ordered. After the initial three boats had been built, the RNLI decided not to

▼ 17-002 (later A-501) on exercise off Lyme Regis, where she was named Bob Abbott, powered by twin Mercury outboards. (By courtesy of the RNLI)

◀ A-502 being put through her paces off the RNLI Depot at Poole in 1984 during the RNLI's Open Days, with the Poole inshore lifeboat being one of several lifeboats in attendance for the benefit of supporters. (Nicholas Leach)

pursue the design, as the McLachlan was regarded as a better boat.

However, the completed boats saw operational service at Ramsgate, Eastney, Lyme Regis and Poole. The last of the three, 17-003, later designated A-502, had a sixteen-year career, mostly at Poole. She did not leave service until 1985. At Poole she was initially moored at the Fishermen's Dock, but in 1974 was moved to Lilliput Marina, with the station's all-weather lifeboat, where pontoons were used for crew boarding. The Dell Quay Dory, with its shallow draught, was found to be particularly suitable for operations at Poole, which covered a large natural harbour, and the craft was well liked by the volunteers.

Boston Whaler Outrage

When a new ILB was needed for Poole in the 1980s, as 17-003 was at the end of her operational life, the crew requested another Dell Quay Dory, even though the RNLI stated that an Atlantic rigid-inflatable was the only option. However, the Poole crew persisted, but, just as an order was about to be placed with Dell Quay Productions Ltd, the company was put into receivership. It was then suggested that an American-built Boston Whaler Outrage type boat should be acquired as this was deemed to be an ideal boat for the station's unique requirements. A local firm had taken an agency for these boats at the time, and just received of an Outrage fitted out as a fire boat. This was demonstrated to the RNLI, and as a result an order was placed for a new Boston Whaler for Poole.

Local fundraising for a new ILB had been ongoing since 1981, when

▲ A-513 Sam and Iris Coles, the only Boston Whaler built for the RNLI, on exercise off Poole. (By courtesy of Poole RNLI)

Poole Round Table got involved and donated £3,000 towards the cost of a new boat. In the same year the Mayor of Poole's chosen charity was the RNLI, so with the Mayor's support there were sufficient funds for a new boat. Numbered A-513, the Outrage was fitted out by the RNLI's Cowes Base to make her suitable to meet the demands of Poole Harbour. Providing a one-off craft for a station was a highly unusual move by the RNLI. This was before the lifeboat fleet was being standardised with fewer designs, and supplying a bespoke craft for one station was not repeated.

The RNLI's Boston Whaler Outrage had an overall length of 20ft (6m), a maximum beam of 7ft 2in (2.2m) and a maximum draught (at rest) of just 2ft (0.6m). The console and roll bar were fitted in such a way as to allow the boat to pass under the harbour bridges, even at high water. Although the Boston Whaler was a one-off, many standard components from the Atlantic 21 rigid-inflatable were used on the boat. A towing bollard was fitted aft, and VHF radio, first-aid equipment, and spotlights were carried. Power was provided by twin 60hp Evinrude outboard engines, giving a maximum speed of thirty knots. Two twenty-five-gallon fuel tanks gave the boat an endurance at full speed of almost six hours. The boat was operated by a crew of three.

The Boston Whaler entered service in January 1985 and was formally named Sam and Iris Coles on 8 May 1985 at Poole Yacht Club, by

Councillor Peter J. Coles, in memory of his parents. Sam Coles had been Poole station's honorary secretary from 1955 to 1969. After a busy first year at Poole, Sam and Iris Coles went back to Cowes Base for her first survey when she was found to be overweight. It was discovered that, when the hull had been built in the USA, the builders used an open-celled foam as buoyancy instead of closed-cell foam, and the open cells had soaked up large amounts of water. As the hull was still under guarantee, the builders agreed to bring over a new hull with the correct foam buoyancy and the RNLI then transferred all the equipment to the new hull and experienced few further problems.

Sam and Iris Coles served at Poole for almost a decade, being launched on service about once a week, mostly for routine tasks, such as towing in broken-down or stranded boats, helping sailboarders and assisting the all-weather lifeboat, but she saved over 100 lives during her career. When replaced in November 1994 by a standard Atlantic 75, she was used as a workboat at the RNLI's Inshore Lifeboat Centre at Cowes, bringing to an end the RNLI's use of rigid-hulled inshore lifeboats.

Boarding boats

Although no more McLachlan ILBs were built after 1975, unusually two further boats to the McLachlan design were completed for use as boarding boats at Spurn Point by the Humber crew. The two boats,

◀ Boarding boat BB-240 at Spurn Point, when she was serving at the Humber station, beneath the davit which was used to launch her. The two boarding boats were fitted with extensive fendering. Initially tyres were used, but a bespoke fendering system was fitted at the RNLI ILC in Cowes. Unlike the McLachlans built as ILBs, which had inboard engines, the boarding boats had outboards, like the wooden hulled prototype McLachlan ILB. (Nicholas Leach)

▶ BB-239 at the RNLI at the RNLI Depot, Poole in August 2009 after being refitted with modified fendering. After being sold out of service, she ended up at Saul Junction in private hands. (Nicholas Leach)

▶ BB-240 at Spurn Point on station at Humber in May 2010, taking the crew to the all-weather lifeboat. The stainless steel hand rails were fitted to assist crew as they transferred between the lifeboat and boarding boat. After being sold by the RNLI in 2015 she went to Hayling Island but was later moved to Coleraine. (Nicholas Leach)

numbered BB-239 and BB-240, were built in 1991 by apprentices at the International Boatbuilding Training College in Lowestoft. Their hulls were carvel-built, mahogany planked, and fully epoxy coated, and they were powered by twin 50hp two stroke Mariner outboard engines. They served at the Humber until 2013, being launched by davit from the Pilot Jetty to take the full-time lifeboat crew to the all-weather lifeboat moored in the estuary. One of them was used during the summer and the other during the winter. Although the two boats were similar in many respects to the McLachlan, they undertook a unique role at the Humber which meant they were different in less obvious respects.

Careers of the A class ILBs

Relatively few A class inshore lifeboats were built, but the Hatch, McLachlans, Dell Quay Dorys and Boston Whaler gave reasonably good service during the 1970s and 1980s. However, the two original Hatch boats were not particularly long-lasting: Hatch 18-01 went into service in 1967 and 18-03 went to Poole in 1967 before going to Torbay in July 1969, but both had been withdrawn by the mid-1970s. The McLachlans enjoyed greater longevity, and the crews at the stations where they served generally liked them. The first McLachlan, 18-02 (later A-503), served first at Eastney (now Portsmouth) and enjoyed a long career as both an ILB and a boarding boat, while the McLachlans at Weston-super-Mare, Oban, Ramsgate, Plymouth and Torbay undertook numerous rescues. Rigid-hulled ILBs were particularly popular with the volunteers at Poole, where the last of the A class, A-513, saw service until well into the 1990s, by when she was the only such ILB in the RNLI fleet.

▼ McLachlans A-504 and A-507 (Relief) at Weston-super-Mare in February 1973. The former was returning from refit, while the latter departed to take up duties at Peel. (Grahame Farr, by courtesy of the RNLI)

A-1 (ex 18-01)

Key data

BUILT 1966, William Osborne, Littlehampton

STATIONS Falmouth 1967; Shoreham 1967; Plymouth 7.1968 – 5.1972

RECORD 28 launches, 17 lives saved

DISPOSAL Sold out of service in 8.1972 for £600 to R.G. Carter, Kingfishers, Tacket Wood, Kingsbridge, Devon

Service history

18-01 was used for extensive trials and spent time at both Falmouth and Shoreham Harbour. She was displayed at St Malo, France, during International LB Conference, 5-9 June 1967. In 1968 she was extensively modified and fitted with buoyancy blocks to make her self-righting. She then went to Plymouth for four years of service and was used as a boarding boat as well as an inshore lifeboat, covering Plymouth Sound. At some point the buoyancy blocks were removed (see page 13).

▲ 18-01 on trials in the mid-1960s, in the River Arun at Littlehampton. The two Hatch boats and the McLachlans were all built at Littlehampton by William Osborne. (By courtesy of the RNLI)

▲ 18-01 on trials at Littlehampton, shortly after completion by William Osborne. (By courtesy of the RNLI)

▲ 18-01 at Plymouth, after being fitted with a small cockpit towards the stern. The buoyancy blocks, which gave her a self-righting capability (see p.13), were removed for operational service. (By courtesy of the RNLI)

A-2 (ex 18-03)

Key data

BUILT 1966, William Osborne, Littlehampton

STATIONS Poole 7.1967 – 1969; Torbay 7.1969 – 10.1975

RECORD 159 launches, 29 lives saved (at Torbay 1970-75)

DISPOSAL Scrapped 1974

Service history

18-03 arrived at Poole on 22 July 1967 and went into service on 26 July. She was moored alongside the station's offshore lifeboat, the 35ft 6in Liverpool motor Bassett-Green (ON.891) at Poole Quay, for an initial three-month trial. Two years later she was moved to Torbay, and performed her first service call there on 6 July 1969, when the sailing dinghy Muriel capsized off Fishcombe. Upon her arrival, it was found that the dinghy had been righted and that the crew was safe. The dinghy and crew were towed into Brixham by another vessel and no useful service was undertaken by the ILB.

▲ 18-03 (later A-2) at speed in Poole harbour. During her short time at Poole, she was operated alongside the D class inflatable D-69. (By courtesy of the RNLI)

▲ A-2 moored at Brixham while on station at Torbay in the early 1970s, with the 52ft Barnett Princess Alexandra of Kent (ON.945) at moorings in the background. (By courtesy of the RNLI)

▲ A-2 at Torbay, where she was on station for just over five years. (By courtesy of the RNLI)

A-500 (ex 17-01)

Key data

BUILT 1968,Dell Quay Productions Ltd at Itchenor, Sussex

STATIONS Eastney Aug 1968 – Jun 1969, Ramsgate Jul 1969 – Dec 1971, Poole Feb 1972 – Dec 1973

RECORD Eastney: service information unknown; Ramsgate: 23 launches, 16 lives saved; Poole: 21 launches, 7 lives saved

DISPOSAL Sold 1974

Service history

A-500 was the first Dell Quay Dory to enter service with the RNLI, and was the first lifeboat to be built from glass reinforced plastic. She was originally numbered 17-01 (the other Dell Quay boats were -002 and -003) but was redesignated A-500. She answered her first call at Eastney on 27 August 1968 to a capsized sailing dinghy. However, she was deemed to be not well suited to the seas off Eastney so, in June 1969, was withdrawn and, after a routine overhaul, transferred to Ramsgate. After two years at the Kent station, she was moved again, this time to Poole, where she spent almost two years.

▲ 17-01 on trials in August 1968. Her RNLI career lasted only five years. (By courtesy of the RNLI)

▲ 17-01 being put through her paces for the benefit of the media in 1968. (By courtesy of the RNLI)

▲ 17-01 was renumbered A-500 in the early 1970s, when she was on station at Poole, where she is pictured on 30 June 1973, towards the end of her service career. (Jeff Morris)

A-501 (ex 17-002)

Key data

BUILT 1969, Dell Quay Productions Ltd at Itchenor, Sussex

STATIONS Lyme Regis 20.4.1969 – 31.8.1973; Poole 10.7.1974 – 19.5.1975

RECORD 51 launches, 49 lives saved (Lyme Regis); 26 launches, 7 lives saved (Poole)

DISPOSAL Sold out of service 1975

Service history

A-501 was named Bob Abbott while at Lyme Regis, in memory of Robert West Abbott, who had been Coxswain of the station's pulling lifeboats from 1903 to 1925. She initially operated alongside a standard D class inflatable ILB, but this was withdrawn in 1970. A-501, somewhat unusually, was launched from a trolley, which was hauled up a short ramp into the boathouse in the harbour. This had been built in 1967 when the Lyme Regis station was reopened. A-501 was replaced at Lyme Regis by an Atlantic 21 in 1973, being then transferred to Poole and subsequently sold out of service.

▲ 17-002 in service at Lyme Regis, during which time she was named Bob Abbott. On board are Lionel Fisher, Dave Hallett, Albert Hodder (helm), Dr Bowles and Peter Gill. (David Manners, courtesy of Richard Horobin)

◀ 17-002 (A-501) being recovered by trolley into the boathouse at Lyme Regis, fitted with Penta outboard engines. These engines were manufactured by Volvo, who took over Penta. (By courtesy of the RNLI)

◀ 17-002 (A-501), named Bob Abbott, on her launching trolley inside the boathouse at Lyme Regis in the late 1960s. (By courtesy of the RNLI)

◀ A-501 (formerly 17-002) moored at Poole in the mid-1970s, with her twin Mercury outboard engines raised. (Jeff Morris)

A-502 (ex 17-003)

Key data

BUILT 1969, Dell Quay Productions Ltd at Itchenor, Sussex

DONOR Gift of Mrs Zoe Jacobs

STATIONS Poole 6.1969 – 2.1972; Ramsgate 2.1972 – 5.1975; Poole 8.1975 – 1.1985

RECORD Poole: 303 launches, 57 lives saved; Ramsgate: 28 launches, 7 lives saved

DISPOSAL Sold 1985

Service history

17-003 arrived at Poole on 23 June 1969, the day before the Poole station celebrated its centenary. The station also had a D class inflatable on station, but the crew preferred the Dell Quay Dory so she remained as the station's inshore rescue boat.

SERVICES IN SUMMARY

Year • Launches/saved

1970 • 13 / 3 Poole
1971 • 13 / 16 Poole
1972 • 2 / 0 Poole
1972 • 10 / 2 Ramsgate
1973 • 8 / 1 Ramsgate
1974 • 7 / 4 Ramsgate
1975 • 11 / 1 Poole
1975 • 3 / 0 Ramsgate
1976 • 11 / 3 Poole
1977 • 1 / 0 On passage
1977 • 21 / 1 Poole
1978 • 21 / 3 Poole
1979 • 26 / 5 Poole
1980 • 22 / 0 Poole
1981 • 30 / 2 Poole
1982 • 40 / 7 Poole
1983 • 51 / 9 Poole
1984 • 41 / 7 Poole
1985 • 1 / 0 Poole
TOTAL • 332 / 64

▲ 17-003 on station at Poole, moored alongside the 35ft 6in Liverpool class lifeboat George Elmy (ON.873).

▲ A-502 moored alongside at Lilliput Marina, Poole, September 1980. (Tony Denton)

▲ A-502 in Poole harbour, approaching the RNLI Depot, July 1984. (Nicholas Leach)

A-503 (ex 18-02)

Key data

BUILT 1967, William Osborne, Littlehampton

STATIONS Eastney Jul 1967 – Jul 1970, Weston-super-Mare 1971; Pill July 1971 – Oct 1972; Falmouth 1973 – Mar 1980; Humber (boarding boat)

RECORD Eastney: 7 launches, 6 lives saved; Weston: 5 launches, 0 lives saved; Pill: 25 launches, 6 lives saved; Falmouth: 45 launches

DISPOSAL Not known

Service history

18-02, the prototype McLachlan built with a wooden hull, was used for trials initially, after which she entered operational service, being renumbered A-503 in the early 1970s. She was initially powered by a single engine, but later had twin 62hp outboards and then twin inboard engines. Although she was only designated as a rescue asset at Falmouth towards the end of her time there, A-503 undertook forty-five service launches. After being replaced by A-508 at Falmouth, she was refitted at Harrisons boatyard at Amble and was then used as a relief boarding boat at the Humber during the 1980s.

▲ 18-02 on trials shortly after she was completed, prior to being redesignated A-503. (By courtesy of the RNLI)

◀ A-503 on duty at Pill, being put through her paces for the local Carnival Day, 20 May 1972. She was the first lifeboat to serve the new station at Pill, which covered the upper reaches of the Bristol Channel. (Grahame Farr)

◀ A-503 on exercise off Falmouth, where she was designated as a boarding boat as well as a rescue craft. (By courtesy of the RNLI)

◀ A-503 at Spurn Point operating as a boarding boat for the Humber lifeboat. The station was the only one in the RNLI network to be operated by a full-time crew. (By courtesy of Richie Leonard)

A-504 (ex 18-004)

Key data

BUILT 1970, William Osborne, Littlehampton

STATIONS Weston-super-Mare 9.5.1970 – 5.1983

RECORD 176 launches, 60 lives saved

DISPOSAL Displayed at lifeboat museums in Bristol, Cromer and Chatham, being part of the Lifeboat Collection at Chatham Historic Dockyard since the 1990s.

Service history

A-504 had a fine service career at Weston-super-Mare, where she was operated from the lifeboat house and, somewhat unusually, launched from the slipway, down which she was lowered on a bespoke cradle. The D class inflatable was kept on a trolley in front of her.

SERVICES IN SUMMARY
Year • Launches/saved

1970 • 9 / 2	1974 • 22 / 20	1978 • 6 / 0	1982 • 19 / 5
1971 • 19 / 0	1975 • 17 / 7	1979 • 9 / 2	1983 • 1 / 0
1972 • 13 / 6	1976 • 14 / 0	1980 • 10 / 0	TOTAL • 176 / 60
1973 • 13 / 10	1977 • 10 / 8	1981 • 14 / 0	

▲ 18-004, later A-504, on trials off the Sussex coast shortly after being built. (By courtesy of the RNLI)

▲ A-504 on her cradle on the slipway at Weston-super-Mare. The wheeled trolley to the left, beneath the McLachlan's bow, was for the D class inflatable. (Grahame Farr, by courtesy of the RNLI)

▲ A-504 on display as part of the Lifeboat Collection at Chatham Historic Dockyard. (Nicholas Leach)

A-505 (ex 18-005)

Key data

BUILT 1970, William Osborne, Littlehampton

STATIONS Eastney Jul 1970 – Sep 1971, Oban May 1972 – May 1973, Eastney Dec 1974 – Oct 1975, Humber (boarding boat) 1977 – 1987

RECORD Eastney: 30 launches, 6 lives saved; Oban: 12 launches, 6 lives saved

DISPOSAL Sold 4.1.1988 from Leggett's Boatyard, Grimsby

Service history

A-505 had a relatively limited service career, operating most of the time at Eastney (now Portsmouth), as well as completing a short stint at Oban. The station she served for the longest was Humber, where she was operated as a boarding boat at Spurn Point. After being sold out of service, she ended up at R.W. Davis's yard at Saul Junction.

SERVICES IN SUMMARY
Year • Launches/saved

1970 • 14 / 1 Eastney
1971 • 16 / 1 Eastney
1972 • 11 / 6 Oban
1973 • 1 / 0 Oban
1974 • ? / ?
1975 • 10 / 4 Eastney

▲ One of the McLachlans built as part of the first production run at William Osborne's boatyard, 18-005 photographed while she was on trials off the Sussex coast in 1970.

◀ 18-005 on trials. (By courtesy of the RNLI)

◀ A-505 in use as a boarding boat at Spurn Point for the Humber station, June 1982. (Tony Denton)

◀ A-505 was sold out of service in 1988 and in 2018 was undergoing restoration at R.W. Davis's boatyard at Saul Junction, Gloucester. (Nicholas Leach)

A-506 (ex 18-006)

Key data

BUILT 1971, William Osborne, Littlehampton

STATIONS Peel May 1972 – Mar 1973, Relief Mar 1973 – Jun 1976, Poole Dec 1976 – Aug 1979, Plymouth Aug 1979 – Aug 1980

RECORD Peel: 3 launches, 3 lives saved; Poole: 1 launch, 0 lives saved; Plymouth: 20 launches, 0 lives saved

DISPOSAL Unknown; reportedly was to be transferred to the National Lifeboat Museum, Bristol, August 1980.

Service history

A-506 was completed in 1972 and sent to Peel, being placed on station on 20 June 1972 after a period of crew familiarisation. She was launched from a bespoke launching trolley with the Case LA tractor T54, previously used to launch the station's offshore lifeboat. After her stint at Peel, A-506 later served at Poole and Plymouth, where she was kept afloat.

SERVICES IN SUMMARY

Year • Launches/saved

1972 • 2 / 3 Peel
1976 • 1 / 0 Peel
1976 • 1 / 0 Poole
1979 • 9 / 0 Plymouth
1980 • 11 / 0 Plymouth

▲ 18-006, later A-506, on trials off the Sussex coast shortly after being built.

▲ A-506 on station at Plymouth in the late 1970s.

▲ A-506 at the RNLI Depot, Poole in October 1978.

A-507 (ex 18-007)

Key data

BUILT 1971, William Osborne, Littlehampton

STATIONS Weston-super-Mare Dec 1972 – Feb 1973, Peel May 1973 – Aug 1975, Relief 1975 – 1976, Peel Jul – Aug 1976, Plymouth 1977 – Oct 1983

RECORD Weston: 5 launches, 0 lives saved; Peel: 15 launches, 8 lives saved; Plymouth: 58 launches, 12 lives saved

DISPOSAL Sold out of service 1989

Service history

A-507 was displayed at the London Boat Show in 1970 after she had been completed, and was then placed in the Relief Fleet. She had a service career lasting just over ten years, serving at three stations, as well as a short stint in the Relief Fleet, which included time at Ramsgate in 1977. She was in stored at William Osborne's in 1973 and at Dorset Lake Boatyard in 1976, in between her station operational duties.

SERVICES IN SUMMARY

Year • Launches/saved • Station

1972 • 2 / 0 Weston	1975 • 6 / 4	1978 • 12 / 8	1981 • 14 / 2
1973 • 1 / 0	1976 • 2 / 1	1978 • 2 / 0 Weston	1982 • 11 / 1
1973 • 3 / 1 Peel	1977 • 1 / 0 Ramsgate	1979 • 6 / 0 Plymouth	1983 • 2 / 0
1974 • 4 / 2	1977 • 9 / 1 Plymouth	1980 • 4 / 0	

▲ A-507 moored at Plymouth during the late 1970s. (By courtesy of the RNLI)

▲ A-507 on station at Plymouth in July 1982. (Tony Denton)

▲ A-507 undergoing maintenance work at Mashford Bros boatyard, Plymouth.

A-508 (ex 18-008)

Key data

BUILT 1971, William Osborne, Littlehampton

STATIONS Eastney Sep 1971 – Dec 1974, Invergordon July 1976 – Nov 1979, Falmouth Mar 1980 – Jan 1988

RECORD Eastney: 41 launches, 26 lives saved; Invergordon: 1 launch; Falmouth: 92 launches, 19 lives saved

DISPOSAL Sold out of service in 1988

Service history

A-508 had a varied service career, spending three years at Eastney (a station which was later renamed Portsmouth) before a quiet stint at Invergordon in north-east Scotland, followed by eight years at Falmouth, alongside the 52ft Arun Elizabeth Ann (ON.1058). While she was in service at Eastney, A-508 was involved in some outstanding rescues, the first on 18 June 1972, when she went to the aid of the yacht Florya, which had run aground in a south-westerly gale, gusting up to force nine, with very rough seas. Crewed by three experienced lifeboat men, the ILB found the casualty was rolling heavily and pounding on the sands. She rescued all three people on board, landing them at Eastney after a very rough passage. For excellent seamanship, helm John Fletcher was accorded the Thanks of the Institution Inscribed on Vellum, with Vellum Service Certificates being presented to the two crew.

The other outstanding service was undertaken on 5 August 1973, when A-508 went to the assistance of the yacht Jo of Ryde. A helicopter from HMS Daedalus had been unable to lift off the two occupants, because of injuries that they had sustained. In near gale-force south-westerly winds and extremely heavy confused seas, A-508 headed out over the Bar, becoming airborne twice as she battled her way through huge waves. The ILB was manoeuvred to within 15ft of the stern of the yacht and, after two attempts, managed to rescue the two yachtsmen. As one man had a suspected fractured rib and the other had injured his foot, the actual rescue was fraught with difficulties, but with superb skill and teamwork, the lifeboat's two crew succeeded. In winds gusting to force nine and very rough confused seas, the ILB partially broached to and shipped a large amount of water, but this quickly drained away and Helmsman Faro skilfully negotiated the harbour entrance. For truly outstanding seamanship and great courage, Helmsman Dennis Faro was awarded a Bronze Medal, with lifeboat crew member Ronald Needle receiving a Medal Service Certificate.

▲ A-508 moored at Falmouth on 10 July 1984. (Tony Denton)

▲ A-508 at the RNLI Depot, Poole on 26 May 1988 after being replaced at Falmouth.

A-509

Key data

BUILT 1972, William Osborne, Littlehampton

DONOR Gift of Mr and Mrs Dutton-Forshawe, Park Lane, London

STATIONS Plymouth May 1972 – Oct 1976, Brighton May – Jul 1978, Falmouth 1981 – 1983

RECORD Plymouth: 85 launches, 22 lives saved; Brighton: 2 launches, 0 lives saved; Falmouth: 5 launches, 1 life saved

DISPOSAL Sold out of service in June 1988

Service history

A-509 was built for service at Plymouth, where she was used as a boarding boat as well as an inshore lifeboat. After four years at the South Devon station, she had one season at Brighton before going to Falmouth, where she was probably used as a Relief ILB.

SERVICES IN SUMMARY

Year • Launches/saved • Station

1972 • 15 / 12 Plymouth
1973 • 20 / 6 Plymouth
1974 • 12 / 3 Plymouth
1975 • 22 / 1 Plymouth
1976 • 13 / 0 Plymouth
1977 • 3 / 0 Plymouth
1978 • 2 / 0 Brighton
1981 • 2 / 0 Falmouth
1982 • 1 / 1 Falmouth
1983 • 2 / 0 Falmouth

▲ A-509 moored in Millbay Docks at Plymouth, with 44ft Waveney Thomas Forehead and Mary Rowse II (ON.1028) in the background in the mid-1970s. (By courtesy of the RNLI)

▲ A-509 on exercise at Plymouth, where she served for four years. (By courtesy of Plymouth RNLI)

▲ A-509 on exercise when she was stationed at Plymouth. (By courtesy of Plymouth RNLI)

A-510

Key data

BUILT 1975, William Osborne, Littlehampton

STATIONS Pill Jun 1973 – Oct 1974, Ramsgate Jun 1975 – Aug 1984, Boarding boat 1984 – 1988 (at Falmouth and Humber)

RECORD Pill: 14 launches, 10 lives saved; Ramsgate: 141 launches, 32 lives saved; Falmouth: 3 launches, 0 lives saved (in 1985 and 1986)

DISPOSAL Sold in 1989

Service history

A-510 arrived at Pill in June 1973, becoming the second of two McLachlans to serve the station, which had been established in 1971 to cover the upper reaches of the Bristol Channel. The station was closed in 1974, and A-510 was withdrawn, being reallocated to Ramsgate, where she served with distinction for almost a decade.

SERVICES IN SUMMARY

Year • Launches/saved

1973 • 5 / 2 Pill	1977 • 21 / 6	1981 • 16 / 2	1984 • 3 / 0
1974 • 9 / 8	1978 • 19 / 1	1982 • 8 / 4	1985 • 1 / 0
1975 • 14 / 7 Ramsgate	1979 • 10 / 2	1983 • 7 / 2	1986 • 2 / 0
1976 • 22 / 1	1980 • 21 / 7		

▲ A-510 at Pill in May 1974; when she was withdrawn, the station was closed. (Grahame Farr, by courtesy of the RNLI)

▲ A-510 at Ramsgate, returning to harbour. (By courtesy of Clive Lawford)

▲ A-510 at the RNLI Depot, Poole in July 1985, having been withdrawn from Ramsgate. (Nicholas Leach)

A-511

Key data

BUILT 1973, William Osborne, Littlehampton
DONOR Gift of North British Hotels Ltd
STATIONS Oban Jun 1974 – 1982; Relief Apr 1984 – Sep 1985
RECORD 87 launches, 9 lives saved (Oban); 15 launches, 2 lives saved (Falmouth 1984)
DISPOSAL Sold 2 Sep 1985 for £18,500

Service history

A-511 was the penultimate 15ft 6in McLachlan lifeboat to be built, and was sent to Oban as station lifeboat in June 1974, replacing A-505 (see p.46), which had arrived in May 1972 to open the new station on Scotland's west coast. By the time A-511 took over the station was well established. The McLachlans were kept afloat on a mooring in the south-east corner of Oban bay, and a stern line was run ashore. The three crew boarded using a small dinghy. At Oban, the McLachlans were used as lifeboats, rather than boarding boats, and from 1978 A-511 operated alongside 42ft Watson motor lifeboats. Both the Watson and McLachlan were replaced in 1983 by a 33ft Brede class intermediate lifeboat.

SERVICES IN SUMMARY
Year • Launches/saved

1973 • 11/1 Oban
1974 • 10/5
1975 • 8/0
1976 • 9/0
1977 • 13/0
1978 • 9/3
1979 • 6/0
1980 • 9/0
1981 • 11/0
1982 • 1/0
1984 • 15/2 Falmouth

▲ A-511 moored at Oban with the former Moelfre lifeboat Watkin Williams (ON.922). (Courtesy of Oban RNLI)

▲ A-511 on exercise at Oban, where she served for eight years. (By courtesy of Oban RNLI)

▲ A-511 at the RNLI Depot, Poole in May 1988 at the end of her career. (Nicholas Leach)

A-512

Key data

BUILT 1974, William Osborne, Littlehampton
DONOR Gift of the Worshipful Company of Cordwainers
STATIONS Torbay 1975 – 1987; Boarding Boat 1987 – 88
RECORD 176 launches, 55 lives saved
DISPOSAL Scrapped 1988

Service history

A-512 served her entire lifeboat career at Torbay, being moored in Brixham harbour and used as both a boarding boat and a rescue craft. During 1977-78 she was modified and the standard McLachlan helmsman's console was removed, being replaced by two seats further forward; a windscreen and dashboard with engine controls were also fitted.

SERVICES IN SUMMARY

Year • Launches/saved

1975 • 27 / 2
1976 • 25 / 3
1977 • 26 / 10
1978 • 23 / 17
1979 • 11 / 1
1980 • 11 / 9
1981 • 12 / 3
1982 • 15 / 2
1983 • 12 / 3
1984 • 5 / 1
1985 • 3 / 0
1986 • 4 / 4
1987 • 2 / 0

▲ A-512 on station at Torbay, with her original helm configuration. (By courtesy of the RNLI)

▲ A-512 moored in Brixham in 1984, during the latter stages of her time at Torbay. (Nicholas Leach)

▲ A-512 on exercise at Torbay, with the revised helm position.

A-513 Sam and Iris Coles

Key data

BUILT 1985, Boston Whaler, fitted out by the RNLI Inshore Lifeboat Centre, Cowes

STATIONS Poole Jan 1985 – Nov 1994

RECORD 593 launches, 106 lives saved

DISPOSAL Withdrawn 28.11.1994, sent to RNLI ILB Centre, Cowes as a work boat

Service history

SAM AND IRIS COLES (A-513) arrived at Poole in January 1985, a station which had been served by a succession of the earlier Dell Quay Dorys. She was funded by the Mayor of Poole's Charity Appeal and was named Sam and Iris Coles. She capsized on service on 23 November 1986, and was removed from the station for repairs, returning on 10 June 1988. She was withdrawn from the station in late 1994 and replaced by an Atlantic 21.

SERVICES IN SUMMARY
Year • Launches/saved

1985 • 53 / 11
1986 • 38 / 1
1987 • 12 / 0
1988 • 40 / 0
1989 • 41 / 6
1990 • 74 / 28
1991 • 56 / 18
1992 • 75 / 16
1993 • 106 / 12
1994 • 97 / 14
TOTAL • 593 / 106

▲ A-513 in Poole harbour; she saved over 100 lives in almost ten years of service. (By courtesy of the RNLI)

▲ A-513 in Lilliput Marina, Poole, from where she operated until 1989. (By courtesy of John Buckby)

▲ A-513 on exercise in Poole harbour. (By courtesy of the RNLI)

Stations served by A class ILBs

STATIONS SERVED BY THE A CLASS INSHORE LIFEBOATS

(Replacements in brackets)
Eastney (B class Atlantic)
Falmouth (B class Atlantic)
Invergordon (Brede)
Oban (Brede)
Peel (B class Atlantic)
Pill (station closed)
Plymouth (ILB withdrawn)
Poole (B class Atlantic)
Ramsgate (B class Atlantic)
Shoreham Harbour (D class)
Torbay (D class)
Weston-super-Mare (B class Atlantic)